National Professional Resources

MTSS & Mathematics for Middle School

by Jennifer C. Caton & Mary E. Little

The majority of students in the United States consistently perform below proficiency on National Mathematics Assessments (National Assessment of Educational Progress [NAEP], 2019). This number tends to be far worse for students with disabilities, with 74 percent of eighth graders performing below basic when compared to 26 percent of students without disabilities (USDOE, 2019). The Every Student Succeeds Act (ESSA, 2015) focused on implementing Multi-Tiered Systems of Support (MTSS) to advance the education of all struggling students, emphasizing evidence-based instructional strategies. This guide provides critical information, practical strategies, and evidenced-based interventions about middle school mathematics instruction within MTSS. It offers an array of ways to use this highly effective framework to meet the needs of students who struggle with mathematics.

What is Multi-Tiered Systems of Support (MTSS)?

Multi-Tiered Systems of Support (MTSS) is an integrated system of services that is designed to provide students with the instruction and interventions they need to achieve. Using a collaborative, instructional decision-making process and evidence-based practices, teams of educators from different disciplines (e.g., mathematics, special education, language arts, etc.) develop and evaluate instruction and intervention plans with the goal of significantly improving students' performance.

MTSS is operational all day, every day, and for all students, rather than just during a specific period of time or for a specific group of students (Howard, 2009). For example, students who score below expectations on state exams, school district benchmark exams, and/or curriculum-based assessments can receive instruction according to grade–level standards while also receiving additional time re-learning specific math skills (e.g., decimals, fractions, etc.) during another part of the day.

Foundational principles of MTSS as a systematic, dynamic process include:

- Applies to all students, general & special education, English learners (ELs), and gifted;
- Includes all aspects of quality instruction;
- Provides additional supports and services beyond initial instruction;
- Requires continuous progress monitoring and formative assessments of student learning to inform decisions related to instruction and interventions.

Connections: MTSS within State Standards and NCTM Focal Points

Educators within many states have been working together to develop, adopt and implement new standards in mathematics. The Every Student Succeeds Act (ESSA, 2015) focused on providing all children with a fair, equitable, and high-quality education to make progress and close the achievement gap. All educators within schools, local educational agencies, and states are accountable for student academic achievement. Most importantly, it is necessary to focus on student outcomes across the country through quality curriculum, instruction, and teacher preparation.

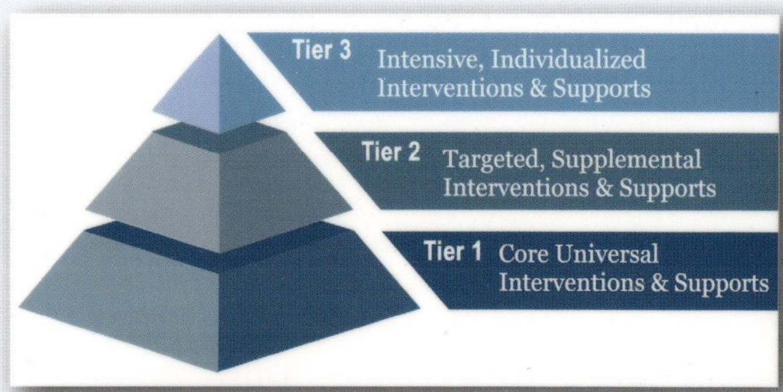

The focus of the revised NCTM standards (2000) is on high-level conceptual understanding and problem-solving, rather than procedural knowledge and rule-driven computation. Teachers must ensure that the instructional goals and standards for conceptual understanding and mathematical proficiency are achieved by each student through high-quality instruction, using evidence-based instructional and intervention methods, products, and practices. Significant research supports the use of MTSS models, including response-to-intervention (RTI), which are widely used in more than 27,000 schools to support student success (Fuchs & Fuchs, 2017; Losinski et al., 2021). The framework and processes within the MTSS model provide support and structure for teachers and students to meet NCTM standards as teams of educators differentiate instruction, provide interventions, and develop conceptual understanding.

The Mathematics Teaching Practices, outlined in *Principles to Actions: Ensuring Mathematical Success for All* (NCTM, 2000), provide an overview of research-informed essential teaching skills that can effectively develop a solid mathematical foundation. These include setting clear goals, implementing high–quality tasks, providing multiple representations, facilitating meaningful discourse, asking purposeful questions, building conceptual understanding, supporting a productive struggle, and using evidence of student thinking. Using this research-informed approach to teaching mathematics ensures that students are engaged in meaningful mathematical learning.

Understanding the MTSS Model

There are three levels, called tiers, within the MTSS model. Each tier uses increasingly more intense instruction and targeted interventions that, together with additional flexibility, differentiation, and scaffolding, help to develop mathematical proficiency among all students based upon their individual learning needs. Data are collected at each tier and used to make instructional decisions about student learning.

Tier 1: Tier 1 instruction occurs within the mathematics classroom and should meet the needs of 70-80% of all learners. This means that all students receive core instruction, with the mathematics teacher using flexible grouping and differentiation within the general education classroom. Screening and progress monitoring are utilized to determine instructional needs and measure student learning. Ongoing professional development is provided to teachers to ensure all students receive quality instruction.

Tier 2: For those students (20-30%) who are not making adequate progress in math, instruction is combined with additional, targeted interventions and supplemental support, depending on the needs of individual students. This level of service is aligned with the curriculum in mathematics. It can be provided by the classroom teacher, special educator, or other school support faculty, such as a mathematics curriculum specialist, instructional coach, or interventionist within the general education classroom. These interventions are delivered in a small group format using strategies that are proven to be effective in addressing these learners. Progress–monitoring data are used to adjust and tailor instruction and intervention for the needs of individual students.

Tier 3: A small percentage of students (5-10%) may still show learning difficulties, even with Tier 2 intervention. They require even more specific support and interventions. Therefore, Tier 3 instruction must be potent and effective (Hall, 2008), and intensified as the duration and frequency increases. Tier 3 services are provided one-on-one and/or in a small group (2-3 students) and are aligned with Tier 1 instruction (universal instructional program) that supports, not replaces, the mathematics curriculum. Tier 3 intervention is provided by highly effective interventionists who are considered "experts" within the area of the intervention. To determine whether instruction and interventions have been successful, teachers and members of the school-based team engage in a problem-solving process. These collaborative, school-based teams of educators meet regularly to review student progress.

Doing What Works:

Assisting Students Struggling with Mathematics: Response to Intervention (RtI) for Elementary and Middle Schools. Visit: https://ies.ed.gov/ncee/wwc/Docs/PracticeGuide/rti_math_pg_042109.pdf

and

From Response to Intervention to Multi-Tiered Systems of Support: Advances in the Science and Practice of Assessment and Intervention. Visit: https://doi.org/10.1007/978-1-4899-7568-3_1

These resources provide excellent information on the MTSS framework, focusing on math. They offer background information, implementation tools, and templates.

Definitions of MTSS Components

TERM	DEFINITION
Continuous Improvement Cycle	Ongoing process of data collection (assessment), through screening, diagnostic, and outcome measures, to inform instruction and identify the appropriate tiers of interventions needed to meet the needs of all students.
Evidence-Based Practices (EBP)	Instruction and intervention practices and programs that have demonstrated improved student results through research.
Fidelity of Implementation	Delivery of instruction and interventions in the manner in which they were designed to be delivered.
Intensity	Increasing support (time, frequency, etc.), explicit and direct instruction, and opportunities for feedback and clarification to improve student learning.
Progress Monitoring	Continuous use of various assessments and probes to measure student academic performance on a regular basis (weekly or monthly), and to adjust and/or intensify instruction and/or interventions, as needed.
Tiers	Three levels of services (instruction and intervention) with increased intensity to meet needs of student(s) within the MTSS framework.

Components of the MTSS Model

The MTSS model consists of the implementation of evidence-based practices (EBP) and interventions in each tier of instruction and intervention, as aligned with national and state standards. Often, school districts purchase these EPBs and/or textbooks for instruction in mathematics and as intervention programs. Adaptations, accommodations, manipulatives, and differentiation strategies are often included within the adopted textbook materials. Additional EBP resources are listed in the What Works Clearinghouse site (https://ies.ed.gov/ncee/WWC). These interventions and instructional approaches are employed with various levels of intensity in terms of time, numbers of students, and so on.

Problem-Solving is used within each tier of instruction and intervention by teams of educators to make decisions. It focuses equal attention on the learner, the environment, and the curriculum. The approach is not linear. The movement throughout the steps is dependent on the student learning and mastery of standards (Little, 2009). This approach includes four steps:

Step # 1: ***Define*** — Determine what the problem is.

Step # 2: ***Analyze*** — Explore the problem using multiple sources of data to determine why it is happening.

Step # 3: ***Implement*** — Establish an achievable goal and develop a plan that includes support to ensure implementation with a high degree of fidelity.

Step # 4: ***Evaluate*** — Assess the success of the intervention based on student data. Determine what needs to be changed to increase student performance and rate of progress.

Academic and Behavior Systems

Tier 1: Core Universal Instruction & Supports
General instruction and support provided to all students in all settings.

Tier 2: Targeted, Supplemental Interventions & Supports
More targeted instruction/intervention and supplemental support in addition to and aligned with the core academic and behavior curriculum.

Tier 3: Intensive, Individualized Interventions & Supports
The most intense (increased time, narrowed focus, reduced group size) instruction and intervention based upon individual student need provided in addition to and aligned with Tier 1 & 2 academic and behavior instruction and supports..

The figure to the right illustrates the continuous nature of the problem-solving model as it is utilized throughout the tiers in the MTSS process. It is a continuous improvement cycle for instruction and intervention in mathematics because the team takes shared responsibility to ensure that instruction and supports are continually effective, starting at the student's initial enrollment in the school through graduation.

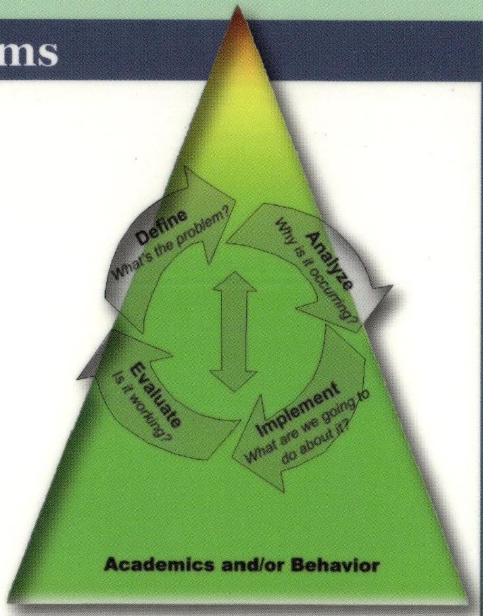

Used with permission from the Florida Department of Education, 2009.

Implementing MTSS in the Mathematics Classroom

The following are specific recommendations for implementing MTSS in the middle school mathematics classroom:

1. Screen all students: (a) to identify those at risk for potential mathematics difficulties and (b) to provide appropriate interventions to students identified as at-risk.
2. Focus instructional materials for students receiving interventions on number concepts and skills, as needed, in grades 4 through 8.
3. Provide explicit and systematic instruction during intervention. This includes providing models of proficient problem-solving, verbalization of thought processes, guided practice, corrective feedback, and frequent cumulative review.
4. Provide instruction on solving word problems that is based on common underlying mathematics proficiencies and structures.
5. Include opportunities for students to work with visual and multiple representations of mathematical ideas.
6. Allot approximately 10 minutes in each lesson to review basic arithmetic facts, if needed.
7. Monitor the progress of students receiving supplemental instruction (Tiers 2 & 3) as well as other students who struggle with mathematics.
8. Include motivational strategies in Tier 2 and 3 interventions.

Throughout implementation of MTSS in mathematics, instruction and interventions are designed and delivered to help middle school students at their individual level of need using differentiated instruction, evidence-based interventions, active engagement methods, and technology. The following chart provides suggestions that target common behaviors.

TARGET STUDENT BEHAVIORS	ACTIVE ENGAGEMENT STRATEGIES	TECHNOLOGY TO USE
Class Participation Exhibits behaviors resulting from a language barrier, and/or lack of engagement	• Pre-teach by providing advance organizer for students • Incorporate cooperative learning strategies • Use graphic organizers • Use manipulatives	• White boards to record/show answers • Signals, charts, software, etc.
Identifying Critical Content Exhibits difficulty understanding essential questions	• Use graphic organizers • Highlight key points • Use visual cue to signal a key point during instruction	• Highlighters or highlighter tape • Video or audio recording
Motivation Exhibits lack of interest in learning, easily discouraged, doesn't complete homework assignments	• Offer points for participation • Emphasize achievements with frequent and specific reinforcing statements • Create opportunities for success	• Software that reviews activities in math • Math-related software programs
Notetaking/Math Journals Exhibits difficulty taking notes and/or identifying important concepts	• Provide copy of teacher's notes • Allow to listen, then copy • Use guided notes • Provide graphic organizers	• AlphaSmart • iPad • Voice recorder
Oral Expression Exhibits difficulty with explaining answers verbally	• Use transparencies • Incorporate white boards • Provide manipulatives	• Board maker • IntelliKeys
Organization Exhibits difficulty in organizing work and/or formulas, or is unable to remember main concepts	• Use reference sheets • Provide enlarged font size • Use mnemonic devices • Shorten length of assignment	• Kidspiration software • Inspiration software
Written Expression Exhibits difficulty with writing	• Support kinesthetic activities (foldables) • Demonstrate through manipulatives • Offer oral presentations	• Recorded responses

Planning for Instruction

Use of Assessment Data

Teachers use assessment data every day to better understand student learning and to plan their instruction. Gathering formative assessment data from many sources builds a complete picture of students' abilities, skills, and dispositions. It helps teachers monitor student learning during instruction to ensure that progress is made in meeting the standards and benchmarks. Multiple sources of data from four types of assessments are used to analyze instruction.

Universal screening assessments collect student performance data that focus on curriculum standards and provide important preliminary or baseline information about students' strengths and weaknesses. Standardized assessments provided at the initiation of a school year or beginning of a unit of study are one source of data. They are generally used by district or school-based administrators as a broad indicator of student knowledge, skills, and abilities for the purpose of instructional planning. Often, middle school teachers use classroom observations, work sample analyses, and curriculum-based assessments in conjunction with the results of these group administered assessments.

Progress monitoring assessments are usually short and frequent skill-based assessments that provide a snapshot of student learning related to the instructional objective. Curriculum-Based Measurement (CBM) is an effective and efficient means of assessing and monitoring students' ongoing progress in the curriculum. These powerful formative assessment tools measure student mastery by assessing engagement, understanding, and progress of student learning related to the curricular goals in mathematics. Formative assessments help guide the needed changes to instruction (e.g., grouping, differentiation, etc.) to ensure that all students are learning the content. They are often school district-developed formative assessments and/or included in adopted textbooks.

Diagnostic assessments identify specific student skills and abilities that should be targeted for intense intervention through aligned, but different, curriculum resources. The results can be used for remediation through adjustment and differentiation of instruction to address students' specific needs. Diagnostic assessments can be administered before and/or during instruction either schoolwide or in an individual classroom setting. For example, a diagnostic test may be administered during the first week of the school year to specifically target multiplication and division of fractions to all 6th grade students. The data are then used to determine specific content needs for determining courses and scheduling. In addition, a teacher may choose to administer a diagnostic assessment to check for student understanding of place value and mastery of conceptual understanding of fractions prior to initiating a unit or as part of determining needed interventions for Tier 2 and Tier 3.

Outcome assessment, also referred to as summative assessment, measures students' comprehensive learning at the end of an established period of time. This occurs at the end of units of study, the end of the school year, and/or when state and/or district assessments are administered. Usually, the purpose of outcome assessments is related to decisions regarding programs for students to meet their instructional needs.

Math Progress-Monitoring Resources and Probes

AIMSweb Progress Monitoring and RTI System: www.aimsweb.com

easyCBM: http://easycbm.com

Intervention Central: www.interventioncentral.org

National Center on Student Progress Monitoring: https://www.progressmonitoring.org/

Schoolhouse Technolgies: https://www.schoolhousetech.com

Voyager Sopris Learning: https://www.voyagersopris.com

The Iris Center: https://iris.peabody.vanderbilt.edu

Types of Aseessment

	Universal Screening	Progress Monitoring/Formative	Diagnostic	Outcome/Summative
Definition	Initial determination of broad-base of student performance	Skill-based, ongoing, and sensitive to small changes in student learning	In-depth, specific information regarding specific knowledge and skills	Final determination of broad-base of student performance
Uses	Broad benchmark attainment index (initial)	Specific academic or behavioral target	Specific academic domain of knowledge, skills, or abilities	Broad benchmark attainment index (final)
Student Focus	Schoolwide	Class or small group	Individual student	Schoolwide
Frequency	Annually/3-4 times per year	Every 3 weeks or weekly	Annually (or as needed) for in-depth evaluation	Annually
Instruction	Class/school and curricular/ program decisions	Effectiveness of instruction and interventions	Selecting appropriate programs and/or educational placements	Align with curriculum goals and instructional planning process
Implication	First step in instruction and intervention planning	Continue and/or revise instruction and/or interventions	Program or curriculum planning	High stakes, based on state and school district policies

Tier 1: Effective Core Instruction in Mathematics

An effective and coherent mathematics program should be guided by a clear set of content standards, and it must be grounded in a clear and shared vision of teaching and learning—the two critical reciprocal actions that link teachers and students and largely determine educational impact (Riccomini & Witzel, 2010). Effective planning and instruction teach the necessary knowledge, skills, and competencies of mathematical understanding and mathematical proficiency. The National Research Council (NRC, 2001) describes five interwoven strands that comprise "mathematical proficiency":

1. **Conceptual Understanding**: understanding mathematic concepts and operations
2. **Procedural Fluency**: accurately and efficiently conducting operations and mathematics problems
3. **Strategic Competence**: formulating/conducting mathematical problems
4. **Adaptive Reasoning**: thinking about, explaining, and justifying mathematical work
5. **Productive Disposition**: appreciating the useful and positive influences of understanding mathematics and how one's disposition towards mathematics influences success.

Differentiate and Intensify Mathematics Instruction and Intervention within MTSS

Strategies	Definitions/Explanations	Examples
Choice Boards	Organizers that contain a variety of activities. Students can choose one or several to complete as they learn a skill or develop a product. Choice boards can be organized requiring students to choose options that focus on several different skills.	Give a choice board that contains a list of possible math activities that students can complete to learn about content goals (e.g., they can choose to complete an inquiry lesson measuring volume using various containers, use a textbook to read about measuring volume, or watch a video in which the steps are explained).
Compacting	A three-step process of adjusting instruction to account for students' prior mastery of learning objectives. It involves: (1) assess to level of knowledge and determine what still needs to be mastered; (2) create plans for what students need to know and excuse students from studying what they already know; and (3) create plans in enriched or accelerated study.	A sixth-grade class is learning to identify the parts of fractions. Diagnostics indicate that two students already know this information. Teach them to add and subtract fractions in a small group. Three other students who have not mastered necessary pre-skills for the unit receive intensive instruction with a math interventionist prior to instruction in fractions.
Flexible Grouping	Students work as part of many different groups depending on the task and/or content. Students may be placed in groups based on readiness or interest and/or learning profile. Group assignments may be purposeful or random. This strategy allows students to work with a wide variety of peers and keeps them from being labeled as advanced or struggling. Teachers can also permit students to select their groups.	Assign groups based on readiness for direct instruction on algebraic concepts, using results of universal screening. A second example is the math teachers within a team (3 or 4) reassign students for a period of time during an instructional unit (4 weeks) based upon the universal screening results for that unit. Another example is the teacher allows students to choose their own groups for projects that investigate famous mathematicians.
Interest Centers/Groups	Settings that provide learning experiences directed toward specific interests of learners. Students choose a topic that is motivating to them.	Focus on specific math skills, such as addition of fractions, and provide relevant activities, such as eating pizza slices. Also, students work in small groups to research a math topic of interest, such as geometry applications to architecture.
Learning Contracts	Agreements between the teacher and the student. The teacher specifies the necessary skills expected to be learned and the required components of the assignment while the student identifies methods for completing the tasks. This strategy: (1) allows students to work at an appropriate pace; (2) can target learning styles; and (3) helps students work independently while learning planning skills and eliminating unnecessary skill practice.	A student agrees to follow a football team over a two-month period and make inferences about players' performances based on their scoring patterns and physical characteristics. Or, a student (with the teacher's guidance) develops a plan for collecting and analyzing the data and conducting and sharing research about baseball statistics.
Tiered Assignments	Assignments designed to instruct students on essential skills. Provided at different levels of complexity, abstractness, and relevance. The curricular content and objective(s) are the same, but the process and/or product are varied according to students' needs (Little, 2012).	In a unit on measurement, some students are taught basic measurement skills, including the use of a ruler to measure the length of objects. Other students can apply measurement skills to problems involving perimeter.

Tier 2: Targeted Interventions

Tier 2 support is temporary in nature and designed to foster and support success from Tier 1. Generally provided for six to eight weeks, Tier 2 interventions are characterized by small-group targeted instruction with greater intensity than Tier 1 differentiated instruction. It is a supplemental approach designed with appropriate materials and evidence-based interventions and practices. Tier 2 services usually occur within the math classroom rather than in a separate setting. The interventions, however, are provided in addition to the mathematics core curriculum. The student who is struggling with mathematics is not removed from his math class for the added supports. Rather, that student remains in the class and receives additional mathematics instruction and support at another time. Therefore, the student receives *more* mathematics instruction rather than separate or different instruction. Data can be used to make decisions about *when* and *what* to provide for specific students. These decisions are constantly adjusted for each learning experience.

Tier 3: Intensive Interventions

Like Tier 2, Tier 3 instructional supports are in addition to the instruction that each student receives. Tier 3 instruction is intensified by increasing the duration and frequency of instruction and intervention. Instruction and intervention should be explicit and systematic, based upon diagnosed student learning needs, using verbalization of thought processes, guided practice, corrective feedback, and frequent cumulative review (Gersten et al, 2009). Services are usually provided one-on-one and/or as small group instruction aligned with the core instructional program for an additional amount of time. Classroom implementation of the intervention plan and continuous progress monitoring are critical at Tier 3. Use ongoing progress monitoring through curriculum-based assessments to determine progress toward goals. The intervention process does not end until the problem (i.e., discrepancy between performance and expectations) is solved and the student is learning the mathematics content successfully.

Collaboration and Scheduling—All 3 Tiers

The key to providing tiered instruction and interventions lies in establishing a workable schedule that maximizes school personnel and the resources of the entire instructional staff or grade-level team. Specific blocks of time each day are devoted to tiered instruction, often referred to as "tier time" or "walk to intervention." By dividing the entire grade-level team of students into tiered instruction, the model provides students who are already achieving at benchmark levels opportunities for enrichment beyond the core instructional program and provides specific interventions for students based upon progress-monitoring data. Interventions, evidence-based resources, and standard protocols are decided on by the school team and district curriculum specialists.

A Sample School Schedule Using Daily Time Blocks for Tiered Instruction

	MONDAY	TUESDAY	WEDNESDAY	THURSDAY	FRIDAY
Period 1	Tier 1 instruction by grade-level classroom teachers	Tier 1 instruction by grade-level classroom teachers	Tier 1 instruction by grade-level classroom teachers	Tier 1 instruction by grade-level classroom teachers	Tier 1 instruction by grade-level classroom teachers
Period 2	Content instruction	Content instruction	Content instruction	Content instruction	Content instruction
Period 3	Core team planning or specials	Tier 3 interventions by grade-level teams	Tier 3 interventions by grade-level teams	Core team planning or specials	Tier 3 interventions by grade-level teams
Period 4	Content instruction or lunch	Content instruction or lunch	Content instruction or lunch	Content instruction or lunch	Content instruction or lunch
Period 5	Tier 1 instruction in mathematics (Schoolwide)	Tier 1 instruction in mathematics (Schoolwide)	Tier 1 instruction in mathematics (Schoolwide)	Tier 1 instruction in mathematics (Schoolwide)	Tier 1 instruction in mathematics (Schoolwide)
Period 6	"Walk to Intervention" (Tier 2 intervention) for literacy or math (Schoolwide)	"Walk to Intervention" (Tier 2 intervention) for literacy or math (Schoolwide)	"Walk to Intervention" (Tier 2 intervention) for literacy or math (Schoolwide)	"Walk to Intervention" (Tier 2 intervention) for literacy or math (Schoolwide)	"Walk to Intervention" (Tier 2 intervention) for literacy or math (Schoolwide)
Period 7	Core team meeting to review progress monitoring	Specials or Tier 3 interventions by grade-level teams	Specials or Tier 3 interventions by grade-level teams	Core team meeting to review progress monitoring for Tier 3 interventions	Specials or Tier 3 interventions by grade-level teams

References & Resources

Every Student Succeeds Act, 20 U.S.C § 6301 (2015). https://www.congress.gov/114/plaws/publ95/PLAW-114publ95.pdf

Fisher, D. & Frey, N. (2012). *Formative assessment in secondary schools* (laminated reference guide). National Professional Resources.

Fuchs, D. & Fuchs, L.S. (2017). Critique of the national evaluation of Response to Intervention: A case for simpler frameworks. *Exceptional Children*, 83(3), 225-268. https://doi.org/10.1177/0014402917693580.

Gersten, R., et al. (2009). *Assisting students struggling with mathematics: Response to Intervention (RtI) for elementary and middle schools* (NCEE 2009-4060). National Center for Education Evaluation and Regional Assistance, Institute of Education Sciences, U.S. Department of Education.

Hall, S. (2008). *Implementing Response to Intervention: A principal's guide*. Corwin Press.

Hosp, J. & M. (2012). *Curriculum-based measurement: A teacher's guide* (laminated reference guide). National Professional Resources.

Hoover, J. & Love, E. (2011). *Supporting school-based response to intervention: A practitioner's model*. Teaching Exceptional Children, 43(3), 40-48.

Howard, M. (2009). *RTI from all sides: What every teacher needs to know*. Heinemann.

Jimerson, S.R., Burns, M.K., VanDerHeyden, A. (2016). From Response to Intervention to Multi-Tiered Systems of Support: Advances in the science and practice of assessment and intervention. *Handbook of Response to Intervention*. Springer. https://doi.org/10.1007/978-1-4899-7568-3_1

Little, M. (2012). *RTI: lesson planning & instructional decision-making*. National Professional Resources.

Little, M. (2009). *Response to Intervention: Classroom instructional decision-making for teachers*. Love.

Losinski, M., Ennis, R.P., Shaw, A., & Gage, N.A. (2021). Supporting students within an MTSS Framework using SRSD fractions: Results of a regression discontinuity design. *Learning Disabilities Research & Practice, 36*(3), 213-223. https://doi.org.10.1111/ldrp.12253.

Love, N. (2011). *Data literacy for teachers* (laminated reference guide). National Professional Resources.

National Council of Teachers of Mathematics. (2000). *Principles and standards for school mathematics*.

Riccomini, P. J. & Witzel, B. S. (2010). *RtI in math*. Corwin Press.

The IRIS Center. (2017). *High-quality mathematics instruction: What teachers should know*. Retrieved from https://iris.peabody.vanderbilt.edu/module/math/

U.S. Department of Education (2019). Institute of Education Sciences, National Center for Education Statistics, National Assessment of Educational Progress (NAEP), 2019 Nation's Report Card. http://nationsreportcard.gov

VanDerHayden, A. (2011). *RTI and math instruction*. RTI Action Network. http://www.rtinetwork.org/learn/what/rtiandmath?tmpl=component

Websites

https://ies.ed.gov/ncee/wwc/practiceguides
https://ies.ed.gov/ncee/wwc/Docs/practiceguide/wwc_mps_tips_072517.pdf
https://ceedar.education.ufl.edu/cems/math/
http://www.rtinetwork.org/middle-school
http://www.mathscore.com/
https://mtss4success.org/
https://www.interventioncentral.org/

Price: $12.95

Order From:

National Professional Resources
1-800-453-7461
www.NPRinc.com
Item # MTMMS

ISBN: 978-1-949961-72-0

51295

9 781949 961720